DARWIN ON EVOLUTION

DARWIN ON EVOLUTION

Words of Wisdom from the Father of Evolution

CHARLES DARWIN

Skyhorse Publishing

Skyhorse Publishing books may be purchased in bulk at special discounts for sales promotion, corporate gifts, fund-raising, or educational purposes. Special editions can also be created to specifications. For details, contact the Special Sales Department, Skyhorse Publishing, 307 West 36th Street, 11th Floor, New York, NY 10018 or info@skyhorsepublishing.com.

Skyhorse® and Skyhorse Publishing® are registered trademarks of Skyhorse Publishing, Inc.®, a Delaware corporation.

Visit our website at www.skyhorsepublishing.com.

10 9 8 7 6 5 4 3 2 1

Library of Congress Cataloging-in-Publication Data is available on file.

Cover design by Jane Sheppard

Print ISBN: 978-1-63220-666-4
Ebook ISBN: 978-1-63220-890-3

Printed in China

CONTENTS

ON THE THEORY AND CONCERNS OF EVOLUTION

A grain in the balance will determine which individual shall live and which shall die—which variety or species shall increase in number, and which shall decrease, or finally become extinct.

—from *On The Origin of Species*

The following proposition seems to me in a high degree probable—namely, that any animal whatever, endowed with well-marked social instincts, the parental and filial affections being here included, would inevitably acquire a moral sense or conscience, as soon as its intellectual powers had become as well, or nearly as well developed, as in man. For, firstly, the social instincts lead an animal to take pleasure in the society of its fellows, to feel a certain amount of sympathy with them, and to perform various services for them.

—from *The Descent of Man*

Intelligence is based on how efficient a species became at doing the things they need to survive.

If it could be demonstrated that any complex organ existed, which could not possibly have been formed by numerous, successive, slight modifications, my theory would absolutely break down. But I can find no such case.

—from *On The Origin of Species*

One general law, leading to the advancement of all organic beings, namely, multiply, vary, let the strongest live and the weakest die.

—from *On The Origin of Species*

I fully agree with all that you say on the advantages of H. Spencer's excellent expression of "the survival of the fittest." This, however, had not occurred to me till reading your letter. It is, however, a great objection to this term that it cannot be used as a substantive governing a verb; and that this is a real objection I infer from

H. Spencer continually using the words, natural selection.

—from a letter to A. R. Wallace July, 1866

A number of species . . . keeping in a body might remain for a long period unchanged, whilst within this same period, several of these species, by migrating into new countries and coming into competition with foreign associates, might become modified; so that we must not overrate the accuracy of organic change as a measure of time. During early periods of the earth's history, when the forms of life were probably fewer and simpler, the rate of change was probably slower; and at the first dawn of life, when very few forms of the simplest structure existed, the rate of change may have been slow in an extreme degree. The whole history of the world, as at present known, although of a length quite incomprehensible by us, will hereafter be recognized as a mere fragment of time, compared with the ages which have elapsed since the first creature, the progenitor of innumerable extinct and living descendants, was created.

Such simple instincts as bees making a beehive could be sufficient to overthrow my whole theory.

Throw up a handful of feathers, and all must fall to the ground according to definite laws; but how simple is this problem compared to the action and reaction of the innumerable plants and animals which have determined, in the course of centuries, the proportional numbers and kinds of trees now growing on the old Indian ruins!

But passing over the endless beautiful adaptations which we everywhere meet with, it may be asked how can the generally beneficent arrangement of the world be accounted for? Some writers indeed are so much impressed with the amount of suffering in the world, that they doubt, if we look to all sentient beings, whether there is more of misery or of happiness; whether the world as a whole is a good or bad one. According to my judgment happiness decidedly prevails, though this would be very difficult to prove. If the truth of this conclusion be granted, it harmonises well with the effects which we might expect from natural selection. If all the individuals of any species were habitually to suffer to an extreme degree, they would neglect to propagate their kind; but we have no reason to believe that this has ever, or at least often occurred. Some other considerations, moreover, lead to the belief that all sentient beings have been formed so as to enjoy, as a general rule, happiness.

—from *The Life and Letters of Charles Darwin*

Many exotic plants have pollen utterly worthless, in the same exact condition as in the most sterile hybrids. When, on the one hand, we see domesticated animals and plants, though often weak and sickly, yet breeding quite freely under confinement; and when, on the other hand, we see individuals, though taken young from a state of nature, perfectly tamed, long-lived, and healthy (of which I could give numerous instances), yet having their reproductive system so seriously affected by unperceived causes as to fail in acting, we need not be surprised at this system, when it does act under confinement, acting not quite regularly, and producing offspring not perfectly like their parents.

We cannot fathom the marvelous complexity of an organic being; but on the hypothesis here advanced this complexity is much increased. Each living creature must be looked at as a microcosm—a little universe, formed of a host of self-propagating organisms, inconceivably minute and as numerous as the stars in heaven.

I will here give a brief sketch of the progress of opinion on the *Origin of Species*. Until recently the great majority of naturalists believed that species were immutable productions, and had been separately created. This view has been ably maintained by many authors. Some few naturalists, on the other hand, have believed that species undergo modification, and that the existing forms of life are the descendants by true generation of pre existing forms. Passing over allusions to the subject in the classical writers (Aristotle, in his "Physicae Auscultationes" (lib.2, cap.8, s.2), after remarking that rain does not fall in order to make the corn grow, any more than it falls to spoil the farmer's corn when threshed out of doors, applies the same argument to organisation; and adds (as translated by Mr. Clair Grece, who first pointed out the passage to me), "So what hinders the different parts (of the body) from having this merely accidental relation in nature? as the teeth, for example, grow by necessity, the front ones sharp, adapted for dividing, and the grinders flat, and serviceable for masticating the food; since they were not made for the sake of this, but it was the result of accident. And in like manner as to other parts in which there appears to exist an adaptation to an end. Wheresoever, therefore, all things together (that is all the parts of one whole) happened like as if they were made for the sake of something, these were preserved, having been appropriately constituted by an internal spontaneity; and whatsoever things were not thus constituted, perished and still perish." We here see the principle of natural selection shadowed forth, but how little Aristotle fully comprehended the principle, is shown by his remarks on the formation of the teeth.), the first author who in modern times has treated it in a scientific spirit was Buffon. But as his opinions fluctuated greatly at different periods, and as he does not enter on the causes or means of the transformation of species, I need not here enter on details."

—from *On The Origin of Species*

As many more individuals of each species are born than can possibly survive; and as, consequently, there is a frequently recurring struggle for existence, it follows that any being, if it vary however slightly in any manner profitable to itself, under the complex and sometimes varying conditions of life, will have a better chance of surviving, and thus be naturally selected. From the strong principle of inheritance, any selected variety will tend to propagate its new and modified form.

—from *On The Origin of Species*

On the other hand, a man may be convinced that God witnesses all his actions, and he may feel deeply conscious of some fault and pray for forgiveness; but this will not, as a lady who is a great blusher believes, ever excite a blush.

The explanation of this difference between the knowledge by God and man of our actions lies, I presume, in man's disapprobation of immoral conduct being somewhat akin in nature to his depreciation of our personal appearance, so that through association both lead to similar results; whereas the disapprobation of God brings up no such association.

But then with me the horrid doubt always arises whether the convictions of man's mind, which has been developed from the mind of the lower animals, are of any value or at all trustworthy.

The struggle for existence inevitably follows from the high geometrical ratio of increase which is common to all organic beings. This high rate of increase is proved by calculation—by the rapid increase of many animals and plants during a succession of peculiar seasons, and when naturalised in new countries. More individuals are born than can possibly survive. A grain in the balance may determine which individuals shall live and which shall die—which variety or species shall increase in number, and which shall decrease, or finally become extinct. As the individuals of the same species come in all respects into the closest competition with each other, the struggle will generally be most severe between them; it will be almost equally severe between the varieties of the same species, and next in severity between the species of the same genus. On the other hand the struggle will often be severe between beings remote in the scale of nature. The slightest advantage in certain individuals, at any age or during any season, over those with which they come into competition, or better adaptation in however slight a degree to the surrounding physical conditions, will, in the long run, turn the balance.

When we compare the individuals of the same variety or sub-variety of our older cultivated plants and animals, one of the first points which strikes us is, that they generally differ more from each other than do the individuals of any one species or variety in a state of nature. And if we reflect on the vast diversity of the plants and animals which have been cultivated, and which have varied during all ages under the most different climates and treatment, we are driven to conclude that this great variability is due to our domestic productions having been raised under conditions of life not so uniform as, and somewhat different from, those to which the parent species had been exposed under nature. There is, also, some probability in the view propounded by Andrew Knight, that this variability may be partly connected with excess of food. It seems clear that organic beings must be exposed during several generations to new conditions to cause any great amount of variation; and that, when the organisation has once begun to vary, it generally continues varying for many generations. No case is on record of a variable organism ceasing to vary under cultivation. Our oldest cultivated plants, such as wheat, still yield new varieties: our oldest domesticated animals are still capable of rapid improvement or modification.

—from *On The Origin of Species*

Frederick Cuvier and several of the older metaphysicians have compared instinct with habit. This comparison gives, I think, an accurate notion of the frame of mind under which an instinctive action is performed, but not necessarily of its origin. How unconsciously many habitual actions are performed, indeed not rarely in direct opposition to our conscious will! Yet they may be modified by the will or reason. Habits easily become associated with other habits, with certain periods of time and states of the body. When once acquired, they often remain constant throughout life. Several other points of resemblance between instincts and habits could be pointed out. As in repeating a well-known song, so in instincts, one action follows another by a sort of rhythm; if a person be interrupted in a song, or in repeating anything by rote, he is generally forced to go back to recover the habitual train of thought: so P. Huber found it was with a caterpillar, which makes a very complicated hammock; for if he took a caterpillar which had completed its hammock up to, say, the sixth stage of construction, and put it into a hammock completed up only to the third stage, the caterpillar simply re-performed the fourth, fifth, and sixth stages of construction. If, however, a caterpillar were taken out of a hammock made up, for instance, to the third stage, and were put into one finished up to the sixth stage, so that much of its work was already done for it, far from deriving any benefit from this, it was much embarrassed, and, in order to complete its hammock, seemed forced to start from the third stage, where it had left off, and thus tried to complete the already finished work.

—from *On the Origin of Species*

I have deeply regretted that I did not proceed far enough at least to understand something of the great leading principles of mathematics, for men thus endowed seem to have an extra sense.

I believe that species come to be tolerably well-defined objects, and do not at any one period present an inextricable chaos of varying and intermediate links: firstly, because new varieties are very slowly formed, for variation is a very slow process, and natural selection can do nothing until favourable variations chance to occur, and until a place in the natural polity of the country can be better filled by some modification of some one or more of its inhabitants. And such new places will depend on slow changes of climate, or on the occasional immigration of new inhabitants, and, probably, in a still more important degree, on some of the old inhabitants becoming slowly modified, with the new forms thus produced and the old ones acting and reacting on each other. So that, in any one region and at any one time, we ought only to see a few species presenting slight modifications of structure in some degree permanent; and this assuredly we do see.

So it will be with the intellectual faculties, since the somewhat abler men in each grade of society succeed rather better than the less able, and consequently increase in number, if not otherwise prevented. When in any nation the standard of intellect and the number of intellectual men have increased, we may expect from the law of the deviation from an average, that prodigies of genius will, as shewn by Mr. Galton, appear somewhat more frequently than before.

ON NATURE

It is difficult to believe in the dreadful but quiet war lurking just below the serene facade of nature.

The affinities of all the beings of the same class have sometimes been represented by a great tree. I believe this simile largely speaks the truth. The green and budding twigs may represent existing species; and those produced during former years may represent the long succession of extinct species. At each period of growth all the growing twigs have tried to branch out on all sides, and to overtop and kill the surrounding twigs and branches, in the same manner as species and groups of species have at all times overmastered other species in the great battle for life. The limbs divided into great branches, and these into lesser and lesser branches, were themselves once, when the tree was young, budding twigs; and this connection of the former and present buds by ramifying branches may well represent the classification of all extinct and living species in groups subordinate to groups. Of the many twigs which flourished when the tree was a mere bush, only two or three, now grown into great branches, yet survive and bear the other branches; so with the species which lived during long-past geological periods, very few have left living and modified descendants. From the first growth of the tree, many a limb and branch has decayed and dropped off; and these fallen branches of various sizes may represent those whole orders, families, and genera which have now no living representatives, and which are known to us only in a fossil state. As we here and there see a thin straggling branch springing from a fork low down in a tree, and which by some chance has been favored and is still alive on its summit, so we occasionally see an animal like the Ornithorhynchus or Lepidosiren, which in some small degree connects by its affinities two large branches of life, and which has apparently been saved from fatal competition by having inhabited a protected station. As buds give rise by growth to fresh buds, and these, if vigorous, branch

out and overtop on all sides many a fee-
bler branch, so by generation I believe it
has been with the great Tree of Life, which
fills with its dead and broken branches
the crust of the earth, and covers the surface with its ever-branching and beautiful
ramifications.

In crossing the Atlantic we hove-to
during the morning of February 16th, close
to the island of St. Paul's. This cluster of
rocks is situated in 0 degs. 58' north latitude,
and 29 degs. 15' west longitude. It is 540
miles distant from the coast of America, and
350 from the island of Fernando Noronha.
The highest point is only fifty feet above the
level of the sea, and the entire circumference
is under three-quarters of a mile. This small
point rises abruptly out of the depths of the
ocean. Its mineralogical constitution is not
simple; in some parts the rock is of a cherty,

in others of a felspathic nature, including thin veins of serpentine. It is a remarkable fact, that all the many small islands, lying far from any continent, in the Pacific, Indian, and Atlantic Oceans, with the exception of the Seychelles and this little point of rock, are, I believe, composed either of coral or of erupted matter. The volcanic nature of these oceanic islands is evidently an extension of that law, and the effect of those same causes, whether chemical or mechanical, from which it results that a vast majority of the volcanoes now in action stand either near sea-coasts or as islands in the midst of the sea.

The climate [in Rio], during the months of May and June, or the beginning of winter, was delightful. The mean temperature, from observations taken at nine o'clock, both morning and evening, was only 72 degs.

It often rained heavily, but the drying southerly winds soon again rendered the walks pleasant. One morning, in the course of six hours, 1.6 inches of rain fell. As this storm passed over the forests which surround the Corcovado, the sound produced by the drops pattering on the countless multitude of leaves was very remarkable, it could be heard at the distance of a quarter of a mile, and was like the rushing of a great body of water. After the hotter days, it was delicious to sit quietly in the garden and watch the evening pass into night. Nature, in these climes, chooses her vocalists from more humble performers than in Europe. A small frog, of the genus Hyla, sits on a blade of grass about an inch above the surface of the water, and sends forth a pleasing chirp: when several are together they sing in harmony on different notes. I had some difficulty in catching a specimen of this frog. The genus Hyla has its toes terminated by small suckers; and I found this animal could crawl up a pane of glass, when placed absolutely perpendicular. Various cicidae and crickets, at the same time, keep up a ceaseless shrill cry, but which, softened by the distance, is not unpleasant. Every evening after dark this great concert commenced; and often have I sat listening to it, until my attention has been drawn away by some curious passing insect.

Man is descended from a hairy, tailed quadruped, probably arboreal in its habits.

The ride was, however, interesting, as the mountain began to show its true form. When we reached the foot of the main ridge, we had much difficulty in finding any water, and we thought we should have been obliged to have passed the night without any. At last we discovered some by looking close to the mountain, for at the distance even of a few hundred yards the streamlets were buried and entirely lost in the friable calcareous stone and loose detritus. I do not think Nature ever made a more solitary, desolate pile of rock;—it well deserves its name of *Hurtado*, or separated. The mountain is steep, extremely rugged, and broken, and so entirely destitute of trees, and even bushes, that we actually could not make a skewer to stretch out our meat over the fire of thistle-stalks. The strange aspect of this mountain is contrasted by the sea-like plain, which not only abuts against its steep sides, but likewise separates the parallel ranges. The uniformity of the colouring gives an extreme quietness to the view,— the whitish grey of the quartz rock, and the light brown of the withered grass of the plain, being unrelieved by any brighter tint. From custom, one expects to see in the neighbourhood of a lofty and bold mountain, a broken country strewed over with huge fragments. Here nature shows that the last movement before the bed of the sea is changed into dry land may sometimes be one of tranquillity.

Animals, whom we have made our slaves, we do not like to consider our equal.

. . . from the war of nature, from famine and death, the most exalted object which we are capable of conceiving, namely, the production of the higher animals, directly follows. There is grandeur in this view of life, with its several powers, having been originally breathed by the Creator into a few forms or into one; and that, whilst this planet has gone cycling on according to the fixed law of gravity, from so simple a beginning endless forms most beautiful and most wonderful have been, and are being, evolved.

Man selects only for his own good: Nature only for that of the being which she tends.

—from *On The Origin of Species*

In the case of every species, many different checks, acting at different periods of life, and during different seasons or years, probably come into play; someone check or some few being generally the most potent, but all concur in determining the average number or even the existence of the species. In some cases it can be shown that widely-different checks act on the same species in different districts. When we look at the plants and bushes clothing an entangled bank, we are tempted to attribute their proportional numbers and kinds to what we call chance. But how false a view is this! Everyone has heard that when an American forest is cut down, a very different vegetation springs up; but it has been observed that ancient Indian ruins in the Southern United States, which must formerly have been cleared of trees, now display the same beautiful diversity and proportion of kinds as in the surrounding virgin forests. What a struggle between the several kinds of trees must here have gone on during long centuries, each annually scattering its seeds by the thousand; what war between insect and insect—between insects, snails, and other animals with birds and beasts of prey—all striving to increase, and all feeding on each other or on the trees or their seeds and seedlings, or on the other plants which first clothed the ground and thus checked the growth of the trees!

Thus, from the war of nature, from famine and death, the most exalted object which we are capable of conceiving, namely, the production of the higher animals, directly follows. There is grandeur in this view of life, with its several powers, having been originally breathed into a few forms or into one; and that, whilst this planet has gone cycling on according to the fixed law of gravity, from so simple a beginning endless forms most beautiful and most wonderful have been, and are being, evolved.

—from *On The Origin of Species*

Probably in no single instance should we know what to do, so as to succeed. It will convince us of our ignorance on the mutual relations of all organic beings; a conviction as necessary, as it seems to be difficult to acquire. All that we can do, is to keep steadily in mind that each [79]organic being is striving to increase at a geometrical ratio; that each at some period of its life, during some season of the year, during each generation or at intervals, has to struggle for life, and to suffer great destruction. When we reflect on this struggle, we may console ourselves with the full belief, that the war of nature is not incessant, that no fear is felt, that death is generally prompt, and that the vigorous, the healthy, and the happy survive and multiply.

I have stated, that in the thirteen species of ground-finches, a nearly perfect gradation may be traced, from a beak extraordinarily thick, to one so fine, that it may be compared to that of a warbler.

—from *Voyage of the Beagle*

The weather is quite delicious. Yesterday, after writing to you, I strolled a little beyond the glade for an hour and a half and enjoyed myself—the fresh yet dark green of the grand Scotch firs, the brown of the catkins of the old birches, with their white stems, and a fringe of distant green from the larches, made an excessively pretty view. At last I fell asleep on the grass, and awoke with a chorus of birds singing around me, and squirrels running up the trees, and some woodpeckers laughing, and it was as pleasant and rural a scene as I ever saw, and I did not care one penny how any of the beasts or birds had been formed.

—from *Charles Darwin's Letters*

But a plant on the edge of a desert is said to struggle for life against the drought, though more properly it should be said to be dependent upon the moisture."

—from *The Origin of Species*

What a book a devil's chaplain might write on the clumsy, wasteful, blundering, low, and horribly cruel work of nature!

But when on shore, & wandering in the sublime forests, surrounded by views more gorgeous than even Claude ever imagined, I enjoy a delight which none but those who have experienced it can understand - If it is to be done, it must be by studying Humboldt."

As this storm passed over the forests which surround the Corcovado, the sound produced by the drops pattering on the countless multitude of leaves was very remarkable, it could be heard at the distance of a quarter of a mile, and was like the rushing of a great body of water. After the hotter days, it was delicious to sit quietly in the garden and watch the evening pass into night. Nature, in these climes, chooses her vocalists from more humble performers than in Europe. A small frog, of the genus Hyla, sits on a blade of grass about an inch above the surface of the water, and sends forth a pleasing chirp: when several are together they sing in harmony on different notes. I had some difficulty in catching a specimen of this frog. The genus Hyla has its toes terminated by small suckers; and I found this animal could crawl up a pane of glass, when placed absolutely perpendicular. Various cicidae and crickets, at the same time, keep up a ceaseless shrill cry, but which, softened by the distance, is not unpleasant. Every

evening after dark this great concert commenced; and often have I sat listening to it,
until my attention has been drawn away by
some curious passing insect.

The earthquake, however, must be to
everyone a most impressive event: the earth,
considered from our earliest childhood as
the type of solidity, has oscillated like a
thin crust beneath our feet; and in seeing
the labored works of man in a moment
overthrown, we feel the insignificance of
his boasted power.

—from *Voyage of the Beagle*

There are several other sources of
enjoyment in a long voyage, which are of
a more reasonable nature. The map of the
world ceases to be a blank; it becomes a pic-
ture full of the most varied and animated

figures. Each part assumes its proper dimensions: continents are not looked at in the light of islands, or islands considered as mere specks, which are, in truth, larger than many kingdoms of Europe. Africa, or North and South America, are well-sounding names, and easily pronounced; but it is not until having sailed for weeks along small portions of their shores, that one is thoroughly convinced what vast spaces on our immense world these names imply."

—from *Voyage of the Beagle*

ON HUMANITY AND THE STRUGGLE FOR EXISTENCE

A man who dares to waste one hour of time has not discovered the value of life.

—from *The Life & Letters of Charles Darwin*

Besides love and sympathy, animals exhibit other qualities connected with the social instincts which in us would be called moral.

This curious and lamentable loss of the higher aesthetic tastes is all the odder, as books on history, biographies, and travels (independently of any scientific facts which they may contain), and essays on all sorts of subjects interest me as much as ever they did. My mind seems to have become a kind of machine for grinding general laws out of large collections of facts, but why this should have caused the atrophy of that part of the brain alone, on which the higher tastes depend, I cannot conceive. A man with a mind more highly organized or better constituted than mine, would not, I suppose, have thus suffered; and if I had to live my life again, I would have made a rule to read some poetry and listen to some music at least once every week; for perhaps the parts of my brain now atrophied would thus have been kept active through use. The loss of these tastes is a loss of happiness, and may possibly be injurious to the intellect, and more probably to the moral character, by enfeebling the emotional part of our nature.

We must, however, acknowledge, as it seems to me, that man with all his noble qualities . . . still bears in his bodily frame the indelible stamp of his lowly origin.

Man tends to increase at a greater rate than his means of subsistence.

We can allow satellites, planets, suns, universe, nay whole systems of universes, to be governed by laws, but the smallest insect, we wish to be created at once by special act.

Animals are so abundant in these countries, that humanity and self-interest are not closely united; therefore I fear it is that the former is here scarcely known. One day, riding in the Pampas with a very respectable "estanciero," my horse, being tired, lagged behind. The man often shouted to me to spur him. When I remonstrated that it was a pity, for the horse was quite exhausted, he cried out, "Why not?—never mind—spur him—it is my horse." I had then some difficulty in making him comprehend that it was for the horse's sake, and not on his account, that I did not choose to use my spurs. He exclaimed, with a look of great surprise, "Ah, Don Carlos, que cosa!" It was clear that such an idea had never before entered his head.

I have tried lately to read Shakespeare, and found it so intolerably dull that it nauseated me.

In October 1838, that is, fifteen months after I had begun my systematic enquiry, I happened to read for amusement "Malthus on Population," and being well prepared to appreciate the struggle for existence which everywhere goes on from long-continued observation of the habits of animals and plants, it at once struck me that under these circumstances favourable variations would tend to be preserved, and unfavourable ones to be destroyed. The result of this would be the formation of new species. Here then I had at last got a theory by which to work; but I was so anxious to avoid prejudice, that I determined not for some time to write even the briefest sketch of it. In June 1842 I first allowed myself the satisfaction of writing a very brief abstract of my theory in pencil in 35 pages; and this was enlarged during the summer of 1844 into one of 230 pages, which I had fairly copied out and still possess.

Once as a very little boy whilst at the day school, or before that time, I acted cruelly, for I beat a puppy, I believe, simply from enjoying the sense of power; but the beating could not have been severe, for the puppy did not howl, of which I feel sure, as the spot was near the house. This act lay heavily on my conscience, as is shown by my remembering the exact spot where the crime was committed. It probably lay all the heavier from my love of dogs being then, and for a long time afterwards, a passion. Dogs seemed to know this, for I was an adept in robbing their love from their masters.

Whilst Man, however well-behaved, at best is but a monkey shaved!

—from *On The Origin of Species*

What wretched doings come from the ardor of fame; the love of truth alone would never make one man attack another bitterly.

The loss of these tastes [for poetry and music] is a loss of happiness, and may possibly be injurious to the intellect, and more probably to the moral character, by enfeebling the emotional part of our nature.

—from *The Autobiography of Charles Darwin, 1809–82*

Nevertheless so profound is our ignorance, and so high our presumption, that we marvel when we hear of the extinction of an organic being; and as we do not see the cause, we invoke cataclysms to desolate the world, or invent laws on the duration of the forms of life!

—from *The Origin of Species*

We stopped looking for monsters under our bed when we realized that they were inside us.

It is always advisable to perceive clearly our ignorance.

—from *The Expression of Emotion in Man and Animals*

In the long history of humankind (and animal kind, too) those who learned to collaborate and improvise most effectively have prevailed.

I have always maintained that, excepting fools, men did not differ much in intellect, only in zeal and hard work; and I still think there is an eminently important difference.

There is no fundamental difference between man and animals in their ability to feel pleasure and pain, happiness, and misery.

. . . I feel most deeply that the whole subject is too profound for the human intellect. A dog might as well speculate on the mind of Newton.—Let each man hope & believe what he can.

As man advances in civilization, and small tribes are united into larger communities, the simplest reason would tell each individual that he ought to extend his social instincts and sympathies to all members of the same nation, though personally unknown to him. This point being once reached, there is only an artificial barrier to prevent his sympathies extending to the men of all nations and races.

—from *The Descent of Man*

Nothing is easier than to admit in words the truth of the universal struggle for life, or more difficult—at least I have found it so—than constantly to bear this conclusion in mind.

—from *The Origin of Species*

Although most of our expressive actions are innate or instinctive, as is admitted by everyone, it is a different question whether we have any instinctive power of recognizing them. This has generally been assumed to be the case; but the assumption has been strongly controverted by M. Lemoine. Monkeys soon learn to distinguish, not only the tones of voice of their masters, but the expression of their faces, as is asserted by a careful observer.Dogs well know the difference between caressing and threatening gestures or tones; and they seem to recognize a compassionate tone. But as far as I can make out, after repeated trials, they do not understand any movement confined to the features, excepting a smile or laugh; and this they appear, at least in some cases, to recognize. This limited amount of knowledge has probably been gained, both by monkeys and dogs, through their associating harsh or kind treatment with our actions; and the knowledge certainly is not instinctive. Children, no doubt, would soon learn the movements of expression in their elders in the same manner as animals learn those of man. Moreover, when a child cries or laughs, he knows in a general manner what he is doing and what he feels; so that a very small exertion of reason would tell him what crying or laughing meant in others. But the question is, do our children acquire their knowledge of expression solely by experience through the power of association and reason?

A man's friendships are one of the best measures of his worth.

Blushing is the most peculiar and most human of all expressions.

—from *The Expression of Emotion in Man and Animals*

Of all expressions, blushing seems to be the most strictly human; yet it is common to all or nearly all the races of man, whether or not any change of color is visible in their skin. The relaxation of the small arteries of the surface, on which blushing depends, seems to have primarily resulted from earnest attention directed to the appearance of our own persons, especially of our faces, aided by habit, inheritance, and the ready flow of nerve-force along accustomed channels; and afterwards to have been extended by the power of association to self-attention directed to moral conduct. It can hardly be doubted that many animals are capable of appreciating beautiful colors and even forms, as is shown by the pains which the individuals of one sex take in displaying their beauty before those of the opposite sex. But it does not seem possible that any animal, until its mental powers had been developed to an equal or nearly equal degree with those of man, would have closely considered and been sensitive about its own personal appearance. Therefore we may conclude that blushing originated at a very late period in the long line of our descent.

How paramount the future is to the present when one is surrounded by children.

The highest possible stage in moral culture is when we recognize that we ought to control our thoughts.

The love for all living creatures is the most noble attribute of man.

A struggle for existence inevitably follows from the high rate at which all organic beings tend to increase. Every being, which during its natural lifetime produces several eggs or seeds, must suffer destruction during some period of its life, and during some season or occasional year, otherwise, on the principle of geometrical increase, its numbers would quickly become so inordinately great that no country could support the product. Hence, as more individuals are produced than can possibly survive, there must in every case be a struggle for existence, either one individual with another of the same species, or with the individuals of distinct species, or with the physical conditions of life. It is the doctrine of Malthus applied with manifold force to the whole animal and vegetable kingdoms; for in this case there [64]can be no artificial increase of food, and no prudential restraint from marriage. Although some species may be now increasing, more or less rapidly, in numbers, all cannot do so, for the world would not hold them.

—from *On The Origin of Species*

If I had my life to live over again, I would have made a rule to read some poetry and listen to some music at least once every week.

—from *The Autobiography of Charles Darwin, 1809–82*

ON RELIGION

The mystery of the beginning of all things is insoluble by us; and I for one must be content to remain an agnostic.

I see no good reasons why the views given in this volume should shock the religious views of anyone.

—from *On The Origin of Species*

We can clearly see how it is that all living and extinct forms can be grouped together within a few great classes; and how the several members of each class are connected together by the most complex and radiating lines of affinities. We shall never, probably, disentangle the inextricable web of the affinities between the members of any one class; but when we have a distinct object in view, and do not look to some unknown plan of creation, we may hope to make sure but slow progress.

To suppose that the eye with all its inimitable contrivances for adjusting the focus to different distances, for admitting different amounts of light, and for the correction of spherical and chromatic aberration, could have been formed by natural selection, seems, I confess, absurd in the highest degree ... The difficulty of believing that a perfect and complex eye could be formed by natural selection, though insuperable by our imagination, should not be considered subversive of the theory.

One hand has surely worked throughout the universe.

—from *Voyage of the Beagle*

. . . But I own that I cannot see as plainly as others do, and as I should wish to do, evidence of design and beneficence on all sides of us. There seems to me too much misery in the world. I cannot persuade myself that a beneficent and omnipotent God would have designedly created the Ichneumonidæ with the express intention of their feeding within the living bodies of Caterpillars, or that a cat should play with mice . . . I feel most deeply that the whole subject is too profound for the human intellect. A dog might as well speculate on the mind of Newton. Let each man hope and believe what he can.

—from *The Life & Letters of Charles Darwin*

"He who believes that each equine species was independently created, will, I presume, assert that each species has been created with a tendency to vary, both under nature and under domestication, in this particular manner, so as often to become striped like other species of the genus; and that each has been created with a strong tendency, when crossed with species inhabiting distant quarters of the world, to produce hybrids resembling in their stripes, not their own parents, but other species of the genus. To admit this view is, as it seems to me, to reject a real for an unreal, or at least for an unknown, cause. It makes the works of God a mere mockery and deception; I would almost as soon believe with the old and ignorant cosmogonists, that fossil shells had never lived, but had been created in stone so as to mock the shells now living on the sea-shore."

I think an Agnostic would be the more correct description of my state of mind. The whole subject (of God) is beyond the scope of man's intellect.

Among the scenes which are deeply impressed on my mind, none exceed in sublimity the primeval forests undefaced by the hand of man; whether those of Brazil, where the powers of Life are predominant, or those of Tierra del Fuego, where Death and Decay prevail. Both are temples filled with the varied productions of the God of Nature:—no one can stand in these solitudes unmoved, and not feel that there is more in man than the mere breath of his body. In calling up images of the past, I find that the plains of Patagonia

The question of whether there exists a Creator and Ruler of the Universe has been answered in the affirmative by some of the highest intellects that have ever existed.

(Reason tells me of the) extreme difficulty or rather impossibility of conceiving this immense and wonderful universe, including man with his capability of looking far backwards and far into futurity, as the result of blind chance or necessity. When thus reflecting I feel compelled to look to a First Cause having an intelligent mind in some degree analogous to that of man; and I deserve to be called a Theist.

—from *The Autobiography of Charles Darwin, 1809–82*

From a letter to Dr. Abbot (November 16, 1871), in Darwin gives more fully his reasons for not feeling competent to write on religious and moral subjects:

"I can say with entire truth that I feel honoured by your request that I should become a contributor to the "Index", and am much obliged for the draft. I fully, also, subscribe to the proposition that it is the duty of every one to spread what he believes to be the truth; and I honour you for doing so, with so much devotion and zeal. But I cannot comply with your request for the following reasons; and excuse me for giving them in some detail, as I should be very sorry to appear in your eyes ungracious. My health is very weak: I NEVER pass 24 hours without many hours of discomfort, when I can do nothing whatever. I have thus, also, lost two whole consecutive months this season. Owing to this weakness, and my head being often giddy, I am unable to master new subjects requiring much thought, and can deal only with old materials. At no time am I a quick thinker or writer: whatever I have done in science has solely been by long pondering, patience and industry.

Now I have never systematically thought much on religion in relation to science, or on morals in relation to society; and without steadily keeping my mind on such subjects for a LONG period, I am really incapable of writing anything worth sending to the Index."

In a letter to a Dutch student:

I am sure you will excuse my writing at length, when I tell you that I have long been much out of health, and am now staying away from my home for rest.

It is impossible to answer your question briefly; and I am not sure that I could do so, even if I wrote at some length. But I may say that the impossibility of conceiving that this grand and wondrous universe, with our conscious selves, arose through chance, seems to me the chief argument for the existence of God; but whether this is an argument of real value, I have never been able to decide. I am aware that if we admit a first cause, the mind still craves to know whence it came, and how it arose. Nor can I overlook the difficulty from the immense amount of suffering through the world. I am, also, induced to defer to a certain extent to the judgment of the many able men who have fully believed in God; but here again I see how poor an argument this is. The safest conclusion seems to me that the whole subject is beyond the scope of man's intellect; but man can do his duty.

It has been said that I speak of natural selection as an active power or Deity; but who objects to an author speaking of the attraction of gravity as ruling the movements of the planets? Every one knows what is meant and is implied by such metaphorical expressions; and they are almost necessary for brevity. So again it is difficult to avoid personifying the word Nature; but I mean by nature, only the aggregate action and product of many natural laws, and by laws the sequence of events as ascertained by us. With a little familiarity such superficial objections will be forgotten.

He who believes that each equine species was independently created, will, I presume, assert that each species has been created with a tendency to vary, both under nature and under domestication, in this particular manner, so as often to become striped like other species of the genus; and that each has been created with a strong tendency, when crossed with species inhabiting distant quarters of the world, to produce hybrids resembling in their stripes, not their own parents, but other species of the genus. To admit this view is, as it seems to me, to reject a real for an unreal, or at least for an unknown, cause. It makes the works of God a mere mockery and deception; I would almost as soon believe with the old and ignorant cosmogonists, that fossil shells had never lived, but had been created in stone so as to mock the shells now living on the sea-shore.

Nothing is easier than to admit in words the truth of the universal struggle for life, or more difficult—at least I found it so—than constantly to bear this conclusion in mind. Yet unless it be thoroughly engrained in the mind, the whole economy of nature, with every fact on distribution, rarity, abundance, extinction, and variation, will be dimly seen or quite misunderstood. We behold the face of nature bright with gladness, we often see superabundance of food; we do not see or we forget that the birds which are idly singing round us mostly live on insects or seeds, and are thus constantly destroying life; or we forget how largely these songsters, or their eggs, or their nestlings, are destroyed by birds and beasts of prey; we do not always bear in mind, that, though food may be now superabundant, it is not so at all seasons of each recurring year.

Even the theologians have almost ceased to pit the plain meaning of Genesis against the no less plain meaning of Nature. Their more candid, or more cautious, representatives have given up dealing with Evolution as if it were a damnable heresy, and have taken refuge in one of two courses. Either they deny that Genesis was meant to teach scientific truth, and thus save the veracity of the record at the expense of its authority; or they expend their energies in devising the cruel ingenuities of the reconciler, and torture texts in the vain hope of making them confess the creed of Science. But when the peine forte et dure is over, the antique sincerity of the venerable sufferer always reasserts itself. Genesis is honest to the core, and professes to be no more than it is, a repository of venerable traditions of unknown origin, claiming no scientific authority and possessing none.

ON SCIENCE

It has often and confidently been asserted, that man's origin can never be known: but ignorance more frequently begets confidence than does knowledge: it is those who know little, and not those who know much, who so positively assert that this or that problem will never be solved by science. The conclusion that man is the co-descendant with other species of some ancient, lower, and extinct form, is not in any degree new.

—from *The Descent of Man*

False facts are highly injurious to the progress of science, for they often endure long; but false views, if supported by some evidence, do little harm, for everyone takes a salutary pleasure in proving their falseness; and when this is done, one path towards error is closed and the road to truth is often at the same time opened.

The same high mental faculties which first led man to believe in unseen spiritual agencies, then in fetishism, polytheism, and ultimately in monotheism, would infallibly lead him, as long as his reasoning powers remained poorly developed, to various strange superstitions and customs.

Many of these are terrible to think of—such as the sacrifice of human beings to a blood-loving god; the trial of innocent persons by the ordeal of poison or fire; witchcraft, etc.—yet it is well occasionally to reflect on these superstitions, for they shew us what an infinite debt of gratitude we owe to the improvement of our reason, to science, and to our accumulated knowledge.

—from *The Descent of Man*

A scientific man ought to have no wishes, no affections—a mere heart of stone.

Freedom of thought is best promoted by the gradual illumination of men's minds which follows from the advance of science.

The author of the *Vestiges of Creation* would, I presume, say that, after a certain unknown number of generations, some bird had given birth to a woodpecker, and some plant to the misseltoe, and that these had been produced perfect as we now see them; but this assumption seems to me to be no explanation, for it leaves the case of the coadaptations of organic beings to each other and to their physical conditions of life, untouched and unexplained.

But no pursuit at Cambridge was followed with nearly so much eagerness or gave me so much pleasure as collecting beetles. It was the mere passion for collecting, for I did not dissect them, and rarely compared their external characters with published descriptions, but got them named anyhow. I will give a proof of my zeal: one day, on tearing off some old bark, I saw two rare beetles, and seized one in each hand; then I saw a third and new kind, which I could not bear to lose, so that I popped the one which I held in my right hand into my mouth. Alas! it ejected some intensely acrid fluid, which burnt my tongue so that I was forced to spit the beetle out, which was lost, as was the third one.

It is, therefore, of the highest importance to gain a clear insight into the means of modification and coadaptation. At the commencement of my observations it seemed to me probable that a careful study of domesticated animals and of cultivated plants would offer the best chance of making out this obscure problem. Nor have I been disappointed; in this and in all other perplexing cases I have invariably found that our knowledge, imperfect though it be, of variation under domestication, afforded the best and safest clue. I may venture to express my conviction of the high value of such studies, although they have been very commonly neglected by naturalists.

—from *On The Origin of the Species*

How odd it is that anyone should not see that all observation must be for or against some view if it is to be of any service!

It is no valid objection that science as yet throws no light on the far higher problem of the essence or origin of life. Who can explain the essence of the attraction of gravity?

—from *On The Origin of the Species*

Looking backwards, I can now perceive how my love for science gradually preponderated over every other taste. During the first two years my old passion for shooting survived in nearly full force, and I shot myself all the birds and animals for my collection; but gradually I gave up my gun more and more, and finally altogether, to my servant, as shooting interfered with my work, more especially with making out the geological structure of a country. I discovered, though unconsciously and insensibly, that the pleasure of observing and reasoning was a much higher one than that of skill and sport. That my mind became developed through my pursuits during the voyage is rendered probable by a remark made by my father, who was the most acute observer whom I ever saw, of a sceptical disposition, and far from being a believer in phrenology; for on first seeing me after the voyage, he turned round to my sisters, and exclaimed, "Why, the shape of his head is quite altered.

A MIXED RECEPTION: LETTERS CONCERNING THE RELEASE OF *ON THE ORIGIN OF SPECIES*

"Often a cold shudder has run through me, and I have asked myself whether I may have not devoted myself to a fantasy."

—Charles Darwin

From Erasmus Darwin (His brother.) to Charles Darwin, November 23rd (1859).

Dear Charles,

I am so much weaker in the head, that I hardly know if I can write, but at all events I will jot down a few things that the Dr. (Dr., afterwards Sir Henry Holland.) has said. He has not read much above half, so as he says he can give no definite conclusion, and it is my private belief he wishes to remain in that state . . . He is evidently in a dreadful state of indecision, and keeps stating that he is not tied down to either view, and that he has always left an escape by the way he has spoken of varieties. I happened to speak of the eye before he had read that part, and it took away his breath— utterly impossible—structure, function, etc., etc., etc., but when he had read it he hummed and hawed, and perhaps it was partly conceivable, and then he fell back on the bones of the ear, which were beyond all probability or conceivability. He mentioned a slight blot, which I also observed, that in speaking of the slave-ants carrying one another, you change the species without giving notice first, and it makes one turn back . . .

. . . For myself I really think it is the most interesting book I ever read, and can only compare it to the first knowledge of chemistry, getting into a new world or rather behind the scenes. To me the geographical distribution, I mean the relation of islands to continents, is the most convincing of the proofs, and the relation of the oldest forms to the existing species. I dare say I don't feel enough the absence of varieties, but then I don't in the least know if everything now living were fossilized whether the paleontologists could distinguish them. In fact the a priori reasoning is so entirely satisfactory to me that if the facts won't fit in, why so much the worse for the facts is my feeling. My ague has left me in such a state of torpidity that I wish I had gone through the process of natural selection.

Yours affectionately, E.A.D.

J.D. HOOKER TO CHARLES DARWIN.
KEW, MONDAY.

Dear Darwin,

You have, I know, been drenched with letters since the publication of your book, and I have hence forborne to add my mite. I hope now that you are well through Edition II., and I have heard that you were flourishing in London. I have not yet got half-through the book, not from want of will, but of time—for it is the very hardest book to read, to full profits, that I ever tried—it is so cram-full of matter and reasoning. I am all the more glad that you have published in this form, for the three volumes, unprefaced by this, would have choked any Naturalist of the nineteenth century, and certainly have softened my brain in the operation of assimilating their contents. I am perfectly tired of marvelling at the wonderful amount of facts you have brought to bear, and your skill in marshalling them and throwing them on the enemy; it is also extremely clear as far as I have gone, but very hard to fully appreciate. Somehow it reads very different from the MS., and I often fancy I must have been very stupid not to have more fully followed it in MS. Lyell told me of his criticisms. I did not appreciate them all, and there are many little matters I hope one day to talk over with you. I saw a highly flattering notice in the "English Churchman," short and not at all entering into discussion, but praising you and your book, and talking patronizingly of the doctrine! . . . Bentham and Henslow will still shake their heads I fancy . . .

Ever yours affectionately,

JOS. D. HOOKER.

CHARLES DARWIN TO J.D. HOOKER.
DOWN, DECEMBER 14TH [1859].

My dear Hooker,

Your approval of my book, for many reasons, gives me intense satisfaction; but I must make some allowance for your kindness and sympathy. Any one with ordinary faculties, if he had PATIENCE enough and plenty of time, could have written my book. You do not know how I admire your and Lyell's generous and unselfish sympathy, I do not believe either of you would have cared so much about your own work. My book, as yet, has been far more successful than I ever even formerly ventured in the wildest day-dreams to anticipate. We shall soon be a good body of working men, and shall have, I am convinced, all young and rising naturalists on our side. I shall be intensely interested to hear whether my book produces any effect on A. Gray; from what I heard at Lyell's, I fancy your correspondence has brought him some way already. I fear that there is no chance of Bentham being staggered. Will he read my book? Has he a copy? I would send him one of the reprints if he has not. Old J.E. Gray (John Edward Gray (1800-1875), was the son of S.F. Gray, author of the "Supplement to the Pharmacopoeia." In 1821 he published in his father's name "The Natural Arrangement of British Plants," one of the earliest works in English on the natural method. In 1824 he became connected with the Natural History Department of the British Museum, and was appointed Keeper of the Zoological collections in 1840. He was the author of "Illustrations of Indian Zoology," "The Knowsley Menagerie," etc., and of innumerable descriptive Zoological papers.), at the British Museum, attacked me in fine style: "You have just reproduced Lamarck's doctrine and nothing else, and here Lyell and others have been attacking him for twenty years, and because YOU (with a sneer and laugh) say the very same thing, they are all coming round; it is the most ridiculous inconsistency, etc., etc."

You must be very glad to be settled in your house, and I hope all the improvements satisfy you. As far as my experience goes, improvements are never perfection. I am very sorry to hear that you are still so very busy, and have so much work. And now for the main purport of my note, which is to ask and beg you and Mrs. Hooker (whom it is really an age since I have seen), and all your children, if you like, to come and spend a week here. It would be a great pleasure to me and to my wife . . . As far as we can see, we shall be at home all the winter; and all times probably would be equally convenient; but if you can, do not put it off very late, as it may slip through. Think of this and persuade Mrs. Hooker, and be a good man and come.

Farewell, my kind and dear friend, Yours affectionately, C. DARWIN.

P.S.—I shall be very curious to hear what you think of my discussion on Classification in Chapter XIII.; I believe Huxley demurs to the whole, and says he has nailed his colours to the mast, and I would sooner die than give up; so that we are in as fine a frame of mind to discuss the point as any two religionists.

Embryology is my pet bit in my book, and, confound my friends, not one has noticed this to me.

A. Sedgwick to Charles Darwin

My dear Darwin,

I write to thank you for your work on the *Origin of Species*. It came, I think, in the latter part of last week; but it MAY have come a few days sooner, and been overlooked among my book-parcels, which often remain unopened when I am lazy or busy with any work before me. So soon as I opened it I began to read it, and I finished it, after many interruptions, on Tuesday. Yesterday I was employed—1st, in preparing for my lecture; 2ndly, in attending a meeting of my brother Fellows to discuss the final propositions of the Parliamentary Commissioners; 3rdly, in lecturing; 4thly, in hearing the conclusion of the discussion and the College reply, whereby, in conformity with my own wishes, we accepted the scheme of the Commissioners; 5thly, in dining with an old friend at Clare College; 6thly, in adjourning to the weekly meeting of the Ray Club, from which I returned at 10 P.M., dog-tired, and hardly able to climb my staircase. Lastly, in looking through the "Times" to see what was going on in the busy world.

I do not state this to fill space (though I believe that Nature does abhor a vacuum), but to prove that my reply and my thanks are sent to you by the earliest leisure I have, though that is but a very contracted opportunity. If I did not think you a good-tempered and truth-loving man, I should not tell you that (spite of the great knowledge, store of facts, capital views of the correlation of the various parts of organic nature, admirable hints about the diffusion, through wide regions of many related organic beings, etc., etc.) I have read your book with more pain than pleasure.

Parts of it I admired greatly, parts I laughed at till my sides were almost sore; other parts I read with absolute sorrow, because I think them utterly false and grievously mischievous. You have DESERTED—after a start in that tra-road of all solid physical truth—the true method of induction, and started us in machinery as wild, I think, as Bishop Wilkins's locomotive that was to sail with us to the moon. Many of your wide conclusions are based upon assumptions which can neither be proved nor disproved, why then express them in the language and arrangement of philosophical induction? As to your grand principle—NATURAL SELECTION—what is it but a secondary consequence of supposed, or known, primary facts! Development is a better word, because more close to the cause of the fact? For you do not deny causation. I call (in the abstract) causation the will of God; and I can prove that

He acts for the good of His creatures. He also acts by laws which we can study and comprehend. Acting by law, and under what is called final causes, comprehends, I think, your whole principle. You write of "natural selection" as if it were done curiously by the selecting agent. 'Tis but a consequence of the presupposed development, and the subsequent battle for life. This view of nature you have stated admirably, though admitted by all naturalists and denied by no one of common sense. We all admit development as a fact of history: but how came it about? Here, in language, and still more in logic, we are point-blank at issue. There is a moral or metaphysical part of nature as well a physical. A man who denies this is deep in the mire of folly. 'Tis the crown and glory of organic science that it DOES through FINAL CAUSE, link material and moral; and yet DOES NOT allow us to mingle them in our first conception of laws, and our classification of such laws, whether we consider one side of nature or the other. You have ignored this link; and, if I do not mistake your meaning, you have done your best in one or two pregnant cases to break it. Were it possible (which, thank God, it is not) to break it, humanity, in my mind, would suffer a damage that might brutalize it, and sink the human race into a lower grade of degradation than any into which it has fallen since its written records tell us of its history. Take the case of the bee-cells. If your development produced the successive modification of the bee and its cells (which no mortal can prove), final cause would stand good as the directing cause under which the successive generations acted and gradually improved. Passages in your book, like that to which I have alluded (and there are others almost as bad), greatly shocked my moral taste. I think, in speculating on organic descent, you OVER-state the evidence of geology; and that you UNDER-state it while you are talking of the broken links of your natural pedigree: but my paper is nearly done, and I must go to my lecture-room. Lastly, then, I greatly dislike the concluding chapter—not as a summary, for in that light it appears good—but I dislike it from the tone of triumphant confidence in which you appeal to the rising generation (in a tone I condemned in the author of the *Vestiges*) and prophesy of things not yet in the womb of time, nor (if we are to trust the accumulated experience of human sense and the inferences of its logic) ever likely to be found anywhere but in the fertile womb of man's imagination. And now to say a word about a son of a monkey and an old friend of yours: I am better, far better, than I was last year. I have been lecturing three days a week (formerly I gave six a week) without much fatigue, but I find by the loss of activity and memory, and of all productive powers, that my bodily frame is sinking slowly towards the earth. But I have visions of the future. They are as much a part of myself as my stomach and my heart, and these visions are to have their antitype in solid fruition of what is best and greatest. But on one condition only—that I humbly accept God's revelation of Himself both in his works and in His word, and do my best to act in conformity with that knowledge which He only can give me, and He only can sustain me in doing. If you and I do all this we shall meet in heaven.

I have written in a hurry, and in a spirit of brotherly love, therefore forgive any sentence you happen to dislike; and believe me, spite of any disagreement in some points of the deepest moral interest, your tru-hearted old friend,

A. SEDGWICK.

CHARLES DARWIN TO ASA GRAY. DOWN, JULY 20TH [1856].

. . . It is not a little egotistical, but I should like to tell you (and I do not THINK I have) how I view my work. Nineteen years (!) ago it occurred to me that whilst otherwise employed on Natural History, I might perhaps do good if I noted any sort of facts bearing on the question of the origin of species, and this I have since been doing. Either species have been independently created, or they have descended from other species, like varieties from one species. I think it can be shown to be probable that man gets his most distinct varieties by preserving such as arise best worth keeping and destroying the others, but I should fill a quire if I were to go on. To be brief, I ASSUME that species arise like our domestic varieties with MUCH extinction; and then test this hypothesis by comparison with as many general and pretty well-established propositions as I can find made out,—in geographical distribution, geological history, affinities, etc., etc. And it seems to me that, SUPPOSING that such hypothesis were to explain such general propositions, we ought, in accordance with the common way of following all sciences, to admit it till some better hypothesis be found out. For to my mind to say that species were created so and so is no scientific explanation, only a reverent way of saying it is so and so. But it is nonsensical trying to show how I try to proceed in the compass of a note. But as an honest man, I must tell you that I have come to the heterodox conclusion that there are no such things as independently created species—that species are only strongly defined varieties. I know that this will make you despise me. I do not much underrate the many HUGE difficulties on this view, but yet it seems to me to explain too much, otherwise inexplicable, to be false. Just to allude to one point in your last note, viz., about species of the same genus GENERALLY having a common or continuous area; if they are actual lineal descendants of one species, this of course would be the case; and the sadly too many exceptions (for me) have to be explained by climatal and geological changes. A fortiori on this view (but on exactly same grounds), all the individuals of the same species should have a continuous distribution. On this latter branch of the subject I have put a chapter together, and Hooker kindly read it over. I thought the exceptions and difficulties were so great that on the whole the balance weighed against my notions, but I was much pleased to find that it seemed to have considerable weight with Hooker, who said he had never been so much staggered about the permanence of species.

I must say one word more in justification (for I feel sure that your tendency will be to despise me and my crotchets), that all my notions about HOW species change are derived from long continued study of the works of (and converse with) agriculturists and horticulturists; and I believe I see my way pretty clearly on the means used by nature to change her species and ADAPT them to the wondrous and exquisitely beautiful contingencies to which every living being is exposed . . .

CHARLES DARWIN TO C. LYELL DOWN, 16TH [JUNE, 1856].

My dear Lyell,

I am going to do the most impudent thing in the world. But my blood gets hot with passion and turns cold alternately at the geological strides, which many of your disciples are taking.

Here, poor Forbes made a continent to [i.e., extending to] North America and another (or the same) to the Gulf weed; Hooker makes one from New Zealand to South America and round the World to Kerguelen Land. Here is Wollaston speaking of Madeira and P. Santo "as the sure

and certain witnesses of a former continent." Here is Woodward writes to me, if
you grant a continent over 200 or 300 miles
of ocean depths (as if that was nothing),
why not extend a continent to every island
in the Pacific and Atlantic Oceans? And all
this within the existence of recent species!
If you do not stop this, if there be a lower
region for the punishment of geologists, I
believe, my great master, you will go there.
Why, your disciples in a slow and creeping manner beat all the old Catastrophists
who ever lived. You will live to be the great
chief of the Catastrophists.

There, I have done myself a great deal
of good, and have exploded my passion.

So my master, forgive me, and believe
me, ever yours, C. DARWIN.

P.S. Don't answer this, I did it to ease myself.

CHARLES DARWIN TO J.D. HOOKER.
DOWN, OCTOBER 13TH [1858].

... I have been a little vexed at myself at having asked you not "to pronounce too strongly against Natural Selection." I am sorry to have bothered you, though I have been much interested by your note in answer. I wrote the sentence without reflection. But the truth is, that I have so accustomed myself, partly from being quizzed by my non-naturalist relations, to expect opposition and even contempt, that I forgot for the moment that you are the one living soul from whom I have constantly received sympathy. Believe [me] that I never forget for even a minute how much assistance I have received from you.

You are quite correct that I never even suspected that my speculations were a "jam-pot" to you; indeed, I thought, until quite lately, that my MS. had produced no effect on you, and this has often staggered me. Nor did I know that you had spoken in general terms about my work to our friends, excepting to dear old Falconer, who some few years ago once told me that I should do more mischief than any ten other naturalists would do good, [and] that I had half spoiled you already! All this is stupid egotistical stuff, and I write it only because you may think me ungrateful for not having valued and understood your sympathy; which God knows is not the case. It is an accursed evil to a man to become so absorbed in any subject as I am in mine.

I was in London yesterday for a few hours with Falconer, and he gave me a magnificent lecture on the age of man. We are not upstarts; we can boast of a pedigree going far back in time coeval with extinct species. He has a grand fact of some large molar tooth in the Trias.

I am quite knocked up, and am going next Monday to revive under Water-cure at Moor Park.

My dear Hooker, yours affectionately, C. DARWIN.

EXCERPT FROM
THE FOUNDATION
OF ON THE ORIGIN
OF THE SPECIES:
TWO ESSAYS
WRITTEN IN 1842
AND 1843

I will now recapitulate the course of this work, more fully with respect to the former parts, and briefly [as to] the latter. In the first chapter we have seen that most, if not all, organic beings, when taken by man out of their natural condition, and bred during several generations, vary; that is variation is partly due to the direct effect of the new external influences, and partly to the indirect effect on the reproductive system rendering the organization of the offspring in some degree plastic. Of the variations thus produced, man when uncivilized naturally preserves the life, and therefore unintentionally breeds from those individuals most useful to him in his different states: when even semi-civilized, he intentionally separates and breeds from such individuals. Every part of the structure seems occasionally to vary in a very slight degree, and the extent to which all kinds of peculiarities in mind and body, when congenital and when slowly acquired either from external influences, from exercise, or from disuse [are inherited], is truly wonderful. When several breeds are once formed, then crossing is the most fertile source of new breeds. Variation must be ruled, of course, by the health of the new race, by the tendency to return to the ancestral forms, and by unknown laws determining the proportional increase and symmetry of the body. The amount of variation, which has been effected under domestication, is quite unknown in the majority of domestic beings.

In the second chapter it was shown that wild organisms undoubtedly vary in some slight degree: and that the kind of variation, though much less in degree, is similar to that of domestic organisms. It is highly

probable that every organic being, if subjected during several generations to new
and varying conditions, would vary. It is
certain that organisms, living in an *isolated*
country which is undergoing geological
changes, must in the course of time be so
subjected to new conditions; moreover
an organism, when by chance transported
into a new station, for instance into an
island, will often be exposed to new conditions, and be surrounded by a new series of
organic beings. If there were no power at
work selecting every slight variation, which
opened new sources of subsistence to a
being thus situated, the effects of crossing,
the chance of death and the constant tendency to reversion to the old parent-form,
would prevent the production of new races.
If there were any selective agency at work,
it seems impossible to assign any limit to
the complexity and beauty of the adaptive

structures, which *might* thus be produced: for certainly the limit of possible variation of organic beings, either in a wild or domestic state, is not known.

It was then shown, from the geometrically increasing tendency of each species to multiply (as evidenced from what we know of mankind and of other animals when favored by circumstances), and from the means of subsistence of each species on an *average* remaining constant, that during some part of the life of each, or during every few generations, there must be a severe struggle for existence; and that less than a grain[1] in the balance will determine which individuals shall live and which perish. In a country, therefore, undergoing changes, and cut off from

---

[1] https://www.gutenberg.org/files/22728/22728-h/ 22728-h.htm

the free immigration of species better adapted to the new station and conditions, it cannot be doubted that there is a most powerful means of selection, *tending* to preserve even the slightest variation, which aided the subsistence or defense of those organic beings, during any part of their whole existence, whose organization had been rendered plastic. Moreover, in animals in which the sexes are distinct, there is a sexual struggle, by which the most vigorous, and consequently the best adapted, will oftener procreate their kind.

A new race thus formed by natural selection would be undistinguishable from a species. For comparing, on the one hand, the several species of a genus, and on the other hand several domestic races from a common stock, we cannot discriminate them by the amount of external difference, but only, first, by domestic races not remaining so constant or being so "true" as species are; and secondly by races always producing fertile offspring when crossed. And it was then shown that a race naturally selected—from the variation being slower—from the selection steadily leading towards the same ends, and from every new slight change in structure being adapted (as is implied by its selection) to the new conditions and being fully exercised, and lastly from the freedom from occasional crosses with other species, would almost necessarily be "truer" than a race selected by ignorant or capricious and short-lived man. With respect to the sterility of species when crossed, it was shown not to be a universal character, and when present to vary in degree: sterility also was shown probably to depend less on external than on constitutional differences. And it was shown that when individual animals and plants are placed under new conditions, they become, without losing their healths, as sterile, in the same manner and to the same degree, as hybrids; and it is therefore conceivable that the cross-bred offspring between two species, having different constitutions, might have its constitution affected in the same peculiar manner as when an individual animal or plant is placed under new conditions. Man in selecting domestic races has little wish and still less power to adapt the whole frame to new conditions; in nature, however, where each species survives by a struggle against other species and external nature, the result must be very different.

Races descending from the same stock were then compared with species of the same genus, and they were found to present some striking analogies. The offspring also of races when crossed, that is mongrels, were compared with the cross-bred offspring of species, that is hybrids, and they were found to resemble each other in all their characters, with the one exception of sterility, and even this, when present, often becomes after some generations variable in degree. The chapter was summed up, and it was shown that no ascertained limit to the amount of variation is known; or could be predicted with due time and changes of condition granted. It was then admitted that although the production of new races, undistinguishable from true species, is probable, we must look to the relations in the past and present geographical distribution of the infinitely numerous beings, by which we are surrounded—to their affinities and to their structure—for any direct evidence.

In the third chapter the inheritable variations in the mental phenomena of domestic and of wild organic beings were considered. It was shown that we are not concerned in this work with the first origin of the leading mental qualities; but that tastes, passions, dispositions, consensual movements, and habits all became, either congenitally or during mature life, modified and were inherited. Several of these modified habits were found to correspond in every essential character with true instincts, and they were found to follow the same laws. Instincts and dispositions &c. are fully as important to the preservation and increase of a species as its corporeal structure; and therefore the natural means of selection would act on and modify them equally with corporeal structures. This being granted, as well as the proposition that mental phenomena are variable, and that the modifications are inheritable, the possibility of the several most complicated instincts being slowly acquired was considered, and it was shown from the very imperfect series in the instincts of the animals now existing, that we are not justified in *prima facie* rejecting a theory of the common descent of allied organisms from the difficulty of imagining the transitional stages in the various now most complicated and wonderful instincts. We were thus led on to consider the same question with respect both to highly complicated organs, and to the aggregate of several such organs, that is individual organic beings; and it was shown, by the same method of taking the existing most imperfect series, that we ought not at once to reject the theory, because we cannot trace the transitional stages in such organs, or conjecture the transitional habits of such individual species.

In the Second Part the direct evidence of allied forms having descended from the same stock was discussed. It was shown that this theory requires a long series of intermediate forms between the species and groups in the same classes—forms not directly intermediate between existing species, but intermediate with a common parent. It was admitted that if even all the preserved fossils and existing species were collected, such a series would be far from being formed; but it was shown that we have not *good* evidence that the oldest known deposits are contemporaneous with the first appearance of living beings; or that the several subsequent formations are nearly consecutive; or that any one formation preserves a nearly perfect fauna of even the hard marine organisms, which lived in that quarter of the world. Consequently, we have no reason to suppose that more than a small fraction of the organisms which have lived at any one period have ever been preserved; and hence that we ought not to expect to discover the fossilized sub-varieties between any two species. On the other hand, the evidence, though extremely imperfect, drawn from fossil remains, as far as it does go, is in favor of such a series of organisms having existed as that required. This want of evidence of the past existence of almost infinitely numerous intermediate forms, is, I conceive, much the weightiest difficulty on the theory of common descent; but I must think that this is due to ignorance necessarily resulting from the imperfection of all geological records.

In the fifth chapter it was shown that new species gradually appear, and that the old ones gradually disappear, from the earth; and this strictly accords with our theory. The extinction of species seems to be preceded by their rarity; and if this be so, no one ought to feel more surprise at a species being exterminated than at its being rare. Every species which is not increasing in number must have its geometrical tendency to increase checked by some agency seldom accurately perceived by us. Each slight increase in the power of this unseen checking agency would cause a corresponding decrease in the average numbers of that species, and the species would become rarer: we feel not the least surprise at one species of a genus being rare and another abundant; why then should we be surprised at its extinction, when we have good reason to believe that this very rarity is its regular precursor and cause.

In the sixth chapter the leading facts in the geographical distribution of organic beings were considered—namely, the dissimilarity in areas widely and effectually separated, of the organic beings being exposed to very similar conditions (as for instance, within the tropical forests of Africa and America, or on the volcanic islands adjoining them). Also the striking similarity and general relations of the inhabitants of the same great continents, conjoined with a lesser degree of dissimilarity in the inhabitants living on opposite sides of the barriers intersecting it—whether or not these opposite sides are exposed to similar conditions. Also the dissimilarity, though in a still lesser degree, in the inhabitants of different islands in the same archipelago, together with their similarity taken as a whole with the inhabitants of the nearest continent, whatever its character may be. Again, the peculiar relations of Alpine floras; the absence of mammifers on the smaller isolated islands; and the comparative fewness of the plants and other organisms on islands with diversified stations; the connection between the possibility of occasional transportal from one country to another, with an affinity, though not identity, of the organic beings inhabiting them. And lastly, the clear and striking relations between the living and the extinct in the same great divisions of the world; which relation, if we look very far backward, seems to die away. These facts, if we bear in mind the geological changes in progress, all simply follow from the proposition of allied organic beings having lineally descended from common parent-stocks. On the theory of independent creations they must remain, though evidently connected together, inexplicable and disconnected.

In the seventh chapter, the relationship or grouping of extinct and recent species;

the appearance and disappearance of groups; the ill-defined objects of the natural classification, not depending on the similarity of organs physiologically important, not being influenced by adaptive or analogical characters, though these often govern the whole economy of the individual, but depending on any character which varies least, and especially on the forms through which the embryo passes, and, as was afterwards shown, on the presence of rudimentary and useless organs. The alliance between the nearest species in *distinct* groups being general and not especial; the close similarity in the rules and objects in classifying domestic races and true species. All these facts were shown to follow on the natural system being a genealogical system.

In the eighth chapter, the unity of structure throughout large groups, in species adapted to the most different lives, and the wonderful metamorphosis (used metaphorically by naturalists) of one part or organ into another, were shown to follow simply on new species being produced by the selection and inheritance of successive *small* changes of structure. The unity of type is wonderfully manifested by the similarity of structure, during the embryonic period, in the species of entire classes. To explain this it was shown that the different races of our domestic animals differ less, during their young state, than when full grown; and consequently, if species are produced like races, the same fact, on a greater scale, might have been expected to hold good with them. This remarkable law of nature was attempted to be explained through establishing, by sundry facts, that slight variations originally appear during all periods of life, and that when inherited they tend to appear at the corresponding period of life; according to these principles, in several species descended from the same parent-stock, their embryos would almost necessarily much more closely resemble each other than they would in their adult state. The importance of these embryonic resemblances, in making out a natural or genealogical classification, thus becomes at once obvious. The occasional greater simplicity of structure in the mature animal than in the embryo; the gradation in complexity of the species in the great classes; the adaptation of the larva of animals to independent powers of existence; the immense difference in certain animals in their larval and mature states, were all shown on the above principles to present no difficulty.

In the [ninth] chapter, the frequent and almost general presence of organs and parts, called by naturalists abortive or rudimentary, which, though formed with exquisite care, are generally absolutely useless [was considered]. [These structures,] though sometimes applied to uses not normal,—which cannot be considered as mere representative parts, for they are sometimes capable of performing their proper function,—which are always best developed, and sometimes only developed, during a very early period of life,—and which are of admitted high importance in classification,—were shown to be simply explicable on our theory of common descent.

WHY DO WE WISH TO REJECT THE THEORY OF COMMON DESCENT?

Thus have many general facts, or laws, been included under one explanation; and the difficulties encountered are those which would naturally result from our acknowledged ignorance. And why should we not admit this theory of descent[2]? Can it be shown that organic beings in a natural state are *all absolutely invariable*? Can it be said that the *limit of variation* or the number of varieties capable of being formed under domestication are known? Can any distinct line be drawn *between a race and a species*? To these three questions we may certainly answer in the negative. As long as species were thought to be divided and defined by an impassable barrier of *sterility*, whilst we were ignorant of geology, and imagined that the *world was of short duration*, and the number of its past inhabitants few, we were justified in assuming individual creations, or in saying with Whewell that the beginnings of all things are hidden from man.

---

[2] https://www.gutenberg.org/files/22728/22728-h/ 22728-h.htm

Why then do we feel so strong an inclination to reject this theory—especially when the actual case of any two species, or even of any two races, is adduced—and one is asked, have these two originally descended from the same parent womb? I believe it is because we are always slow in admitting any great change of which we do not see the intermediate steps. The mind cannot grasp the full meaning of the term of a million or hundred million years, and cannot consequently add up and perceive the full effects of small successive variations accumulated during almost infinitely many generations. The difficulty is the same with that which, with most geologists, it has taken long years to remove, as when Lyell propounded that great valleys[3] were hollowed out by the slow action of the waves of the

---

[3] https://www.gutenberg.org/files/22728/22728-h/ 22728-h.htm

sea. A man may long view a grand precipice without actually believing, though he may not deny it, that thousands of feet in thickness of solid rock once extended over many square miles where the open sea now rolls; without fully believing that the same sea which he sees beating the rock at his feet has been the sole removing power.

Shall we then allow that the three distinct species of rhinoceros which separately inhabit Java and Sumatra and the neighboring mainland of Malacca were created, male and female, out of the inorganic materials of these countries? Without any adequate cause, as far as our reason serves, shall we say that they were merely, from living near each other, created very like each other, so as to form a section of the genus dissimilar from the African section, some of the species of which section inhabit very similar and some very dissimilar stations? Shall we say that without any apparent cause they were created on the same generic type with the ancient woolly rhinoceros of Siberia and of the other species which formerly inhabited the same main division of the world: that they were created, less and less closely related, but still with interbranching affinities, with all the other living and extinct mammalia? That without any apparent adequate cause their short necks should contain the same number of vertebra with the giraffe; that their thick legs should be built on the same plan with those of the antelope, of the mouse, of the hand of the monkey, of the wing of the bat, and of the fin of the porpoise. That in each of these species the second bone of their leg should show clear traces of two bones having been soldered and united into one; that the complicated bones of their head should become intelligible on the supposition of their having been formed of three expanded vertebra; that in the jaws of each when dissected young there should exist small teeth which never come to the surface. That in possessing these useless abortive teeth, and in other characters, these three rhinoceroses in their embryonic state should much more closely resemble other mammalia than they do when mature. And lastly, that in a still earlier period of life, their arteries should run and branch as in a fish, to carry the blood to gills which do not exist. Now these three species of rhinoceros closely resemble each other; more closely than many generally acknowledged races of our domestic animals; these three species if domesticated would almost certainly vary, and races adapted to different ends might be selected out of such variations. In this state they would probably breed together, and their offspring would possibly be quite, and probably in some degree, fertile; and in either case, by continued crossing, one of these specific forms might be absorbed and lost in another. I repeat, shall we then say that a pair, or a gravid female, of each of these three species of rhinoceros, were separately created with deceptive appearances of true relationship, with the stamp of inutility on some parts, and of conversion in other parts, out of the inorganic elements of Java, Sumatra and Malacca? or have they descended, like our domestic races, from the same parent-stock? For my own part I could no more admit the former proposition than I could admit that the planets move in their courses, and that a stone falls to the ground, not through the intervention of the secondary and appointed law of gravity, but from the direct volition of the Creator.

Before concluding it will be well to show, although this has incidentally appeared, how far the theory of common descent can legitimately be extended. If we once admit that two true species of the same genus can have descended from the same parent, it will not be possible to deny that two species of two genera may also have descended from a common stock. For in some families the genera approach almost as closely as species of the same genus; and in some orders, for instance in the monocotyledonous plants, the families run closely into each other. We do not hesitate to assign a common origin to dogs or cabbages, because they are divided into groups analogous to the groups in nature. Many naturalists indeed admit that all groups are artificial; and that they depend entirely on the extinction of intermediate species. Some naturalists, however, affirm that though driven from considering sterility as the characteristic of species, that an entire incapacity to propagate together is the best evidence of the existence of natural genera. Even if we put on one side the undoubted fact that some species of the same genus will not breed together, we cannot possibly admit the above rule, seeing that the grouse and pheasant (considered by some good ornithologists as forming two families), the bull-finch and canary-bird have bred together.

No doubt the more remote two species are from each other, the weaker the arguments become in favor of their common descent. In species of two distinct families the analogy, from the variation of domestic organisms and from the manner of their intermarrying, fails; and the arguments from their geographical distribution quite or almost quite fails. But if we once admit the general principles of this work, as far as a clear unity of type can be made out in groups of species, adapted to play diversified parts in the economy of nature, whether shown in the structure of the embryonic or mature being, and especially if shown by a community of abortive parts, we are legitimately led to admit their community of descent. Naturalists dispute how widely this unity of type extends: most, however, admit that the vertebrata are built on one type; the articulata on another; the mollusca on a third; and the radiata on probably more than one. Plants also appear to fall under three or four great types. On this theory, therefore, all the organisms *yet discovered* are descendants of probably less than ten parent-forms.

CONCLUSION.

My reasons have now been assigned for believing that specific forms are not immutable creations. The terms used by naturalists of affinity, unity of type, adaptive characters, the metamorphosis and abortion of organs, cease to be metaphorical expressions and become intelligible facts. We no longer look at an organic being as a savage does at a ship[4] or other great work of art, as at a thing wholly beyond his comprehension, but as a production that has a history which we may search into. How interesting do all instincts become when we speculate on their origin as hereditary habits, or as slight congenital modifications of former instincts perpetuated by the individuals so characterized having been preserved. When we look at every complex instinct and mechanism as the summing up of a long history of contrivances, each most useful to its possessor, nearly in the same

---

4   https://www.gutenberg.org/files/22728/22728-h/ 22728-h.htm

way as when we look at a great mechanical invention as the summing up of the labor, the experience, the reason, and even the blunders of numerous workmen. How interesting does the geographical distribution of all organic beings, past and present, become as throwing light on the ancient geography of the world. Geology loses glory from the imperfection of its archives, but it gains in the immensity of its subject. There is much grandeur in looking at every existing organic being either as the lineal successor of some form now buried under thousands of feet of solid rock, or as being the co-descendant of that buried form of some more ancient and utterly lost inhabitant of this world. It accords with what we know of the laws impressed by the Creator[5] on matter that the production and

---

[5] https://www.gutenberg.org/files/22728/22728-h/ 22728-h.htm

extinction of forms should, like the birth and death of individuals, be the result of secondary means. It is derogatory that the Creator of countless Universes should have made by individual acts of His will the myriads of creeping parasites and worms, which since the earliest dawn of life have swarmed over the land and in the depths of the ocean. We cease to be astonished[6] that a group of animals should have been formed to lay their eggs in the bowels and flesh of other sensitive beings; that some animals should live by and even delight in cruelty; that animals should be led away by false instincts; that annually there should be an incalculable waste of the pollen, eggs and immature beings; for we see in all this the inevitable consequences of one great law, of the multiplication of organic beings not

---

[6] https://www.gutenberg.org/files/22728/22728-h/ 22728-h.htm

created immutable. From death, famine, and the struggle for existence, we see that the most exalted end which we are capable of conceiving, namely, the creation of the higher animals, has directly proceeded. Doubtless, our first impression is to disbelieve that any secondary law could produce infinitely numerous organic beings, each characterized by the most exquisite workmanship and widely extended adaptations: it at first accords better with our faculties to suppose that each required the fiat of a Creator. There is a grandeur in this view of life with its several powers of growth, reproduction and of sensation, having been originally breathed into matter under a few forms, perhaps into only one, and that whilst this planet has gone cycling onwards according to the fixed laws of gravity and whilst land and water have gone on replacing each other—that from so simple an origin, through the selection of infinitesimal varieties, endless forms most beautiful and most wonderful have been evolved.